THE SCIENCE OF AI

How Does AI Work?

Traci Pedersen

ReferencePoint Press

San Diego, CA

For more information, contact:
ReferencePoint Press, Inc.
PO Box 27779
San Diego, CA 92198
www.ReferencePointPress.com

LIBRARY OF CONGRESS CATALOGING-IN-PUBLICATION DATA

Name: Pedersen, Traci- author.
Title: How Does AI Work?/by Traci Pedersen
Description: San Diego, CA : ReferencePoint Press, Inc, 2027. | Series: The Science of AI | Includes bibliographical references and index.
Identifiers: LCCN 2026010220 (print) | ISBN 9781678213145 library binding | ISBN 9781678213152 ebook
Subjects: LCSH: Juvenile literature

For compete cataloging-in-publication data please go to www.loc.gov.

CONTENTS

Meet AI: The Technology Shaping Our World

A hungry college student opens his refrigerator, hoping to make a quick meal from what he already has. He sees a few eggs, a bell pepper, some spinach, a block of cheese, and some leftover pasta. Unsure of how to combine these ingredients, he types them into an artificial intelligence (AI) tool. Within seconds, the system suggests several options: a baked pasta casserole, a cheesy vegetable omelet, or a stuffed bell pepper. Just a decade ago, finding recipe options would have taken much more time—perhaps paging through cookbooks, searching online, or experimenting through trial and error. Today, AI can quickly analyze possibilities and provide suggestions, making problem solving faster and more convenient.

"Artificial intelligence will reach human levels by around 2029. Follow that out further to, say, 2045, we will have multiplied the intelligence, the human biological machine intelligence of our civilization a billion-fold."[1]

—Ray Kurzweil, computer scientist and futurist

This kind of convenience hints at how powerful AI is becoming. These systems can process enormous amounts of information and generate responses almost instantly. Many experts believe this is only the beginning. Futurist and computer scientist Ray Kurzweil predicts, "Artificial intelligence will reach human levels by around 2029. Follow that out further to, say, 2045, we will have multiplied the intelligence, the human biological machine intelligence of our civilization a billion-fold."[1]

Whether or not this timeline proves exact, AI is already significantly changing how people learn, work, and create. It can generate text, images, and solutions with impressive speed and accuracy, and the natural language of its responses can make it seem almost human. To understand AI's astounding capabilities and limitations, it helps to look beneath the surface at the system that makes it work.

The Hidden Engine Behind AI

Data is the foundation of modern AI. As British mathematician Clive Humby observed more than two decades ago, "Data is the new oil,"[2] highlighting its central role in powering AI and other technologies in the manner that petroleum fueled the automotive and industrial boom. AI systems learn by studying large collections of information, such as text, numbers, images, audio, or other content. AI does not understand this information like humans do; it simply detects patterns and relationships within it.

AI has made life much more convenient. For example, a college student looking in the refrigerator can easily ask AI how to combine any ingredients to make a delicious meal.

This data comes from many sources, including books, websites, apps, sensors, and databases. Humans or automated programs collect this information, convert it into digital formats, and feed it into AI systems during training. For example, books may be turned into machine-readable text, while databases are organized into structured formats such as tables or lists.

Next, algorithms determine what the system does with that data. Algorithms are sets of instructions that function much like a recipe, providing step-by-step directions for the system to follow. The choice of algorithm depends on the specific task the AI is designed to perform. Some algorithms guide AI in recognizing objects in photos, others in predicting the next word in a sentence, and still others in identifying trends in numerical data. Data scientist Cathy O'Neil describes this process in a TED Talk: "I use an algorithm every day to make a meal for my family. The data I use is the ingredients in my kitchen, the time I have, and the ambition I have—and I curate that data."[3] This illustrates how algorithms influence outcomes by determining how data is used.

Finally, training is the process through which AI runs data through its algorithms in hopes of improving outcomes. During training, the AI applies the algorithm to the dataset and compares its outputs to the correct or expected results. If the AI makes a mistake, it adjusts its calculations slightly so that it is more likely to get the correct result next time. This cycle of making predictions, checking them, and adjusting continues many times, often millions or billions of times, until the AI becomes accurate enough to handle new, unpracticed examples with accuracy.

The Versatility of AI

One of the most powerful features of advanced AI is its ability to generalize. This means it can take what it has learned from past examples and apply it to new situations. For example, an AI trained to recognize objects in photos might learn that chairs usually have a seat, legs, and a backrest. When it sees a new chair, even one with an unusual shape, it can still identify it as a chair because it detects these familiar features. Similarly, a language AI can create sentences it has never seen by using patterns it has learned about how words fit together.

> **"[AI programs are] advanced informational tools rather than entities possessing a level of consciousness that would warrant their categorization alongside sentient beings."[5]**
>
> —Matjaž Gams and Sebastjan Kramar, computer scientists

Some of the most advanced AI systems that demonstrate this versatility are called generative models. These models produce new content based on patterns in their training data. Examples include tools such as ChatGPT, which generates text, or DALL-E, which creates images from written descriptions. As AI researcher Ian Goodfellow explains, "Generative models are a key enabler of machine creativity, allowing machines to go beyond what they have seen before and create something new."[4]

AI and the Illusion of Understanding

At first glance, AI can seem like a brilliant, tireless thinker—able to solve problems and make decisions at incredible speed. Its output often appears nuanced and humanlike. However, its apparent intelligence comes entirely from pattern recognition within datasets and analysis processes based on preprogrammed algorithms. It has no capacity for conscious thought or understanding.

Researchers Matjaž Gams and Sebastjan Kramar explain in their 2024 study that text-based AI programs are "advanced informational tools rather than entities possessing a level of consciousness that would warrant their categorization alongside sentient beings."[5] In other words, although AI can produce impressive and convincing outputs, it does not understand or experience what it creates.

To understand how AI reached its current level of sophistication, it helps to look back at the ideas that laid the groundwork for its foundation. Early experiments—along with advances in computer processing, neural networks, and machine learning—have shaped the systems that define AI in the modern world. Examining this history provides important context for understanding both the strengths and limitations of contemporary AI.

CHAPTER ONE

The Evolution of AI

A few decades ago it would have been hard to imagine machines that could answer nuanced questions instantly, write stories on demand, or navigate traffic using live data. Today artificial intelligence allows computers to perform all these tasks and more. Its strength lies in its ability to recognize patterns, learn from experience, and make decisions far faster than any human could.

These advanced capabilities grew out of decades of research, experimentation, and new discoveries about how machines process information. Early AI systems relied on strict rules and logical operations. Over time, these systems evolved into models that improve through practice and adapt to new situations, coming closer to the flexibility of human thought. Some speculate that this machine intelligence will reach—or has already reached—human-level thinking. While this is still debated, such beliefs demonstrate how far artificial intelligence has come in the span of roughly eight decades.

From Idea to Intelligence

Before AI became the powerful tool it is today, it began as a bold idea: Could machines ever learn to think and solve problems like humans? One of the most influential figures in this early work was British mathematician Alan Turing. Turing had an extraordinary ability to think logically and solve complex problems. During World War II he applied this talent to help break the Enigma code, which the German military used to send secret messages. Cracking the code gave the Allies access to critical wartime information.

> **"What we want is a machine that can learn from experience."[6]**
>
> —Alan Turing, mathematician

After the war, Turing laid important groundwork for artificial intelligence. In 1950 he proposed a thought experiment he called the imitation game. In this game, a person interacts with two hidden players—one human and one machine—and tries to determine which is which based only on their responses. If the machine can respond well enough to be indistinguishable from the human, it is said to have passed the test. Today, the imitation game is known as the Turing test, and it remains one of the most famous early concepts in AI.

Turing stressed that the key to intelligent machines was their ability to learn for themselves. "What we want is a machine that can learn from experience," he said during a 1947 lecture on computer intelligence. He added that the "possibility of letting the machine alter its own instructions provides the mechanism for this."[6] In other words, he imagined machines that could change their behavior over time, rather than simply following fixed rules. This idea laid the groundwork for today's AI and inspired years of advancements in computing.

A few decades ago, it would have been hard to imagine a device that could navigate traffic using live data. Today artificial intelligence allows computers to perform this task and many more.

Other early AI pioneers helped expand these ideas. John McCarthy, an American computer scientist, coined the term *artificial intelligence* and organized the 1956 Dartmouth Conference, widely considered the official birth of AI research. He described AI as "the science and engineering of making intelligent machines, especially intelligent computer programs." He further clarified that intelligence means "the computational part of the ability to achieve goals in the world."[7]

Another early pioneer of AI was fellow American mathematician Marvin Minsky. In 1959 he and McCarthy cofounded the Artificial Intelligence Laboratory at the Massachusetts Institute of Technology (MIT). At the lab, Minsky worked on early neural networks—computer systems modeled after the human brain that learn to recognize patterns—and on reasoning machines that explored how computers might solve problems that normally require human intelligence.

Minsky was also known for his bold predictions about how quickly AI might advance. In a 1970 interview with *Life* magazine, he said, "In from three to eight years we will have a machine with the general intelligence of an average human being. . . . If we're lucky, they might decide to keep us as pets."[8] The statement reflected both the excitement and the unease surrounding AI at the time, since many researchers believed human-level machine intelligence was close. Although those predictions proved far too

Marvin Minsky, seen here in 2006, cofounded the Artificial Intelligence Laboratory at MIT. He explored how computers might solve problems that normally require human intelligence.

optimistic, they show how strongly early scientists believed in AI's potential.

First Working Systems: Rule-Based AI

In the 1970s and 1980s researchers began building programs that attempted to replicate human-level reasoning. These early programs, known as expert systems, were designed to imitate the decision-making of skilled professionals. One of the most influential examples was MYCIN, a medical program created at Stanford University to help diagnose and treat blood infections.

MYCIN worked through a large collection of if–then rules—statements that connected a specific observation to a likely conclusion. For instance, a rule might say: If a patient's lab test shows a certain type of bacteria and the person has a high fever, then recommend a particular counteractive antibiotic. Using hundreds of these rules, MYCIN could interpret symptoms, ask for additional details, suggest tests, and eventually offer a diagnosis and treatment plan. In controlled evaluations at Stanford Medical School, it reached a diagnosis accuracy of around 70 percent, performing about as well as infectious-disease specialists and often better than general practitioners.

Still, its creators were careful not to overstate what the system could do. As Professor Bruce G. Buchanan, who worked on MYCIN, explains, "These are programs with very narrow limits. There's no widespread intelligence here. Yet." He describes the goal more modestly: "Our goal is to build a program that can assist working scientists with reasoning problems. You wouldn't expect a tool to have all of the power of a working scientist, but you would hope that you could design a smart system to provide some of the solutions to subproblems."[9] In other words, MYCIN was meant to act as a support tool—handling smaller, well-defined tasks—rather than replacing a doctor's full judgment or expertise.

Even with their strengths, rule-based systems had clear limits. They depended on fixed rules written by humans, which meant they struggled with unusual cases, conflicting information, or situations that did not fit neatly into an if–then structure. These challenges showed researchers that AI would eventually need a more flexible approach—one that could learn, adapt, and improve on its own.

Patterns from Data: The Rise of Machine Learning

This shift led researchers toward a new idea: machine learning, which became the foundation of modern AI. Instead of following fixed, handwritten rules, computers could learn directly from data. By studying many examples, computer programs identified patterns in the data that helped improve their output. With more data, these computer models could subsequently make better predictions and adapt what they learned from one dataset and apply it to another. For example, in the 1950s and 1960s, scientists began using computers that learned from examples, rather than fixed rules, to recognize and classify handwritten alphabet characters, a process that would later be used to read addresses and sort mail at post offices. This type of approach allowed AI systems to handle problems that were far too complex to program step-by-step.

As machine learning developed, researchers explored several different predictive models. In the 1980s Israeli American computer scientist Judea Pearl introduced Bayesian networks, a type of machine learning model that uses probability to show how pieces of information are connected. Instead of making all-or-nothing decisions, these networks allow computers to weigh uncertainty and update their predictions as new data appears. Pearl later described their impact this way: "With Bayesian networks, we had taught machines to think in shades of grey, and this was an important step toward humanlike thinking."[10] Bayesian networks are still used today, especially in fields like medicine, risk analysis, and decision-making systems, though they are not the main approach behind today's large AI models.

Around the same time, neural networks were gaining renewed attention. Inspired loosely by the structure of the human brain,

Why MYCIN Never Reached Real Hospitals

Even though MYCIN, a system developed in the 1970s, showed impressive accuracy in tests, it was never used on real patients. Many doctors were uncomfortable relying on a computer for life-or-death decisions. Even if MYCIN performed well in studies, physicians preferred their own judgment, especially during emergencies.

There were also legal concerns. If a patient was harmed because a doctor followed MYCIN's advice, no one knew who would be responsible—the doctor, the hospital, or the software creators. At the time, there were no rules or guidelines for using AI in medicine, which made hospitals hesitant.

Technology was another barrier. MYCIN ran on a large mainframe computer, long before electronic health records or modern hospital computer networks. To use it, a doctor had to sit at a separate terminal and type in every detail about a patient. This process was slow and did not match the fast pace of real medical work.

Finally, MYCIN could only handle a very narrow group of infections. It had no commonsense reasoning and could not deal with cases outside its specialty. Together, these obstacles kept MYCIN in the lab, but its programming and capabilities helped shape future medical AI systems.

neural networks use layers of connected nodes to detect patterns in data. The nodes can be thought of as small processing units that examine information, perform calculations when appropriate, and pass on the output to other nodes. Early versions of these networks were limited, but improvements in computing power and larger datasets eventually made them far more effective. Over time, neural networks evolved into deep learning, which now powers many modern AI systems, including image recognition, speech assistants, and large language models.

Machine Learning in Genomics

During the 1990s and early 2000s, machine learning continued to grow and improve. Networks became more powerful and flexible. They could handle unexpected situations and improve as they received more data. Around this time, researchers began applying machine learning to biology. Early systems helped identify genes in DNA and understand their function, laying the groundwork for studying how genes affect health and disease.

These successes set the stage for the powerful genomic AI systems that exist today. A recent example is Evo 2, released in February 2025 by researchers at the Arc Institute in collaboration with Nvidia and several California universities. Evo 2 is trained on trillions of DNA sequences from thousands of species, including bacteria, plants, animals, and humans. Unlike earlier systems, it does not rely on human-written rules about DNA. Instead, it learns patterns directly from DNA sequences, treating DNA like a language and learning the "grammar" of life.

Patrick Hsu, cofounder of the Arc Institute and co–senior author on the Evo 2 research paper, explains, "Evo 2 has a generalist understanding of the tree of life that is useful for a multitude of tasks, from predicting disease-causing mutations to designing potential code for artificial life."[11] This modern AI can work with many types of genomes, helping scientists find harmful mutations, understand gene function, and even design new sequences for research or medicine. In this way, Evo 2 represents a natural extension of decades of machine learning in genomics, connecting the past to today's cutting edge research.

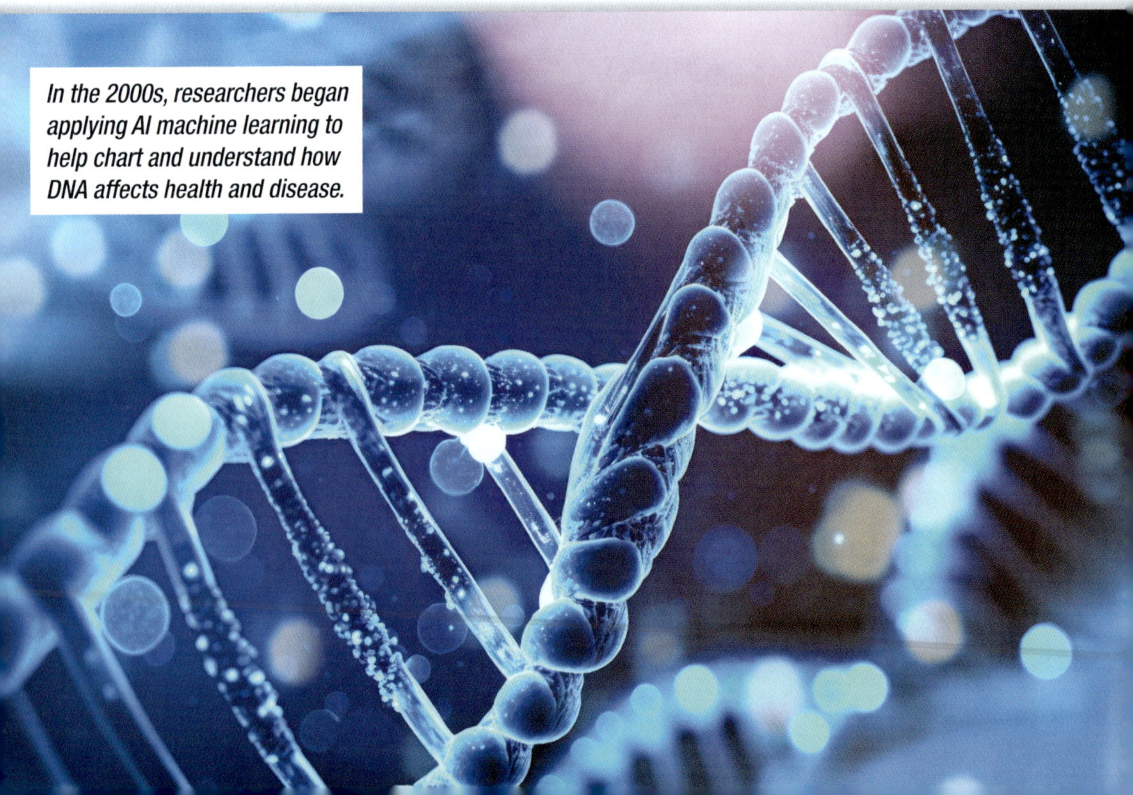

In the 2000s, researchers began applying AI machine learning to help chart and understand how DNA affects health and disease.

Neural Networks in Action at the Post Office

Reading handwritten zip codes was one of the first big, real-world uses of neural networks. In the 1980s and 1990s, the US Postal Service handled millions of letters every day. Having humans read every address took a lot of time and effort, especially because people wrote numbers in many different ways. Neural networks offered a faster and more reliable solution. They became one of the first tools that could read handwritten zip codes and process large amounts of mail efficiently.

Proof of this success came in 1989. Yann LeCun, a researcher at AT&T Bell Labs, and his colleagues trained a neural network on 7,291 scans of handwritten zip codes provided by the US Postal Service. They then tested it on 2,007 additional scans. The network correctly recognized the zip codes 95% of the time. This showed that machines could handle the variety of handwriting accurately across a huge volume of mail.

Genomics is just one example of how machine learning has become central to modern AI. "In just the last five or 10 years, machine learning has become a critical way, arguably the most important way, most parts of AI are done," says MIT Sloan professor Thomas W. Malone. "So that's why some people use the terms AI and machine learning almost as synonymous . . . most of the current advances in AI have involved machine learning."[12]

How Transformers Changed AI

While Evo 2 and genomic AI showed what was possible in specialized applications, work on general AI continued to advance. By the early 2010s, faster computers and better learning methods allowed AI to tackle more complex tasks than ever before. However, even at this stage, models often struggled with language. They had trouble following long texts, recognizing nuances and subtleties, and interpreting ideas arising from the interconnection of multiple sentences. Researchers realized that new approaches were needed to help AI process information more clearly and handle a wider range of tasks.

> "In just the last five or 10 years, machine learning has become a critical way, arguably the most important way, most parts of AI are done."[12]
>
> —Thomas W. Malone, MIT Sloan professor

To address these challenges, researchers developed a new type of neural network called a transformer. This design helps AI

"Learning always wins. The history of AI reflects the reality that it always works better to have a model learn something for itself rather than have a human hand-engineer it."[13]

—Llion Jones, software engineer

understand language much more accurately than older systems. Traditional models had to read text in strict order, one word at a time, which made it hard for computers to connect information within or across sentences. Transformers solved this problem using self-attention, a method that lets the model look at all the words in a sentence at once and figure out how they relate to each other. By analyzing many factors in parallel instead of sequentially, transformers can infer context from the patterns within those relationships. Training these models on existing texts further strengthens these learned connections, enabling them to follow longer explanations, notice subtle details, and track ideas across multiple paragraphs. In short, the model learns how to recognize and interpret rather complex linguistic patterns.

The transformer design was first described in a 2017 research paper titled "Attention Is All You Need," which became one of the most influential papers in modern AI. Llion Jones, a British software engineer and coauthor of the paper, emphasizes, "Learning always wins. The history of AI reflects the reality that it always works better to have a model learn something for itself rather than have a human hand-engineer it."[13] In many ways this echoes Alan Turing's vision decades ago—machines that can learn from experience and adjust their own instructions rather than just follow fixed rules.

With transformers in place, researchers developed large language models, or LLMs. An LLM is a very large neural network trained on massive collections of text. Instead of memorizing what it reads, it learns patterns in language—how sentences are structured, how ideas connect, and how people communicate. OpenAI's ChatGPT and Google's Gemini are examples of LLMs that were built using transformers. Because they learn from vast amounts of information, these models can answer questions, explain difficult topics, and write in many styles. These advances explain why modern AI feels natural, flexible, and helpful in everyday life. And they hint at how next-generation approaches promise to take AI far beyond the capabilities of current transformer models.

CHAPTER TWO

How AI Learns

Learning is what sets artificial intelligence apart from traditional computer programs. Early computers could only follow fixed rules written by humans, producing the same results every time. Modern AI systems, by contrast, are designed to learn from data. They can recognize patterns, improve with experience, and respond to situations they were not explicitly programmed to handle.

This ability to learn is what allows AI to understand language, recognize images, and make predictions about the world. By studying how AI learns—through data, feedback, and repeated practice—it becomes clear why today's systems are more flexible and powerful than earlier forms of computing.

Data: The Fuel Behind AI Learning

"There is no AI without data" is a common saying in the AI field, and it shows a basic truth: AI cannot think or learn on its own. It only improves by studying information, and the more high-quality data it has, the better it can find patterns and do tasks well.

AI usually starts with data provided by humans. When a new system is created, engineers and researchers collect large sets of information—like text, photos, audio, or video—and feed it to the AI to begin learning. Humans also decide where this data comes from, such as databases, websites, or sensors that automatically gather information. One example of a dataset used to train AI is the Pile, a large, openly available collection of text data created by the nonprofit group EleutherAI. The Pile includes billions of words from books, academic papers, and online content, providing a wide-ranging foundation for language models to learn patterns in written text.

Humans continue to help AI as it learns. Trained data labelers tag images, text, or other data with identifiers to help in sorting. Researchers and developers select datasets and set learning goals. Everyday users can also give indirect feedback, such as rating an answer on ChatGPT. These actions show the AI what works and what does not, helping it improve over time.

After the AI is trained, it can also collect new data automatically. Apps and devices record user activity, cameras capture images, and sensors gather information from the environment. AI does not experience the world directly—it only processes the digital information it receives.

Sometimes real-world data is rare or hard to get, such as unusual traffic events or sensitive medical records. In these situations, researchers turn to synthetic data—artificially created information that mimics real data patterns without containing private details. This approach is especially useful in health care because patient privacy is critical. "In healthcare, synthetic data can supplement real datasets, providing a wider range of scenarios for

Computers used to only follow fixed rules written by humans. Modern AI systems now are designed to learn and respond to situations they were not explicitly programmed to handle.

training models, leading to better diagnostic and predictive capabilities,"[14] says Raul Salles de Padua, director of engineering, AI and quantum at Multiverse Computing.

> "Data is going to be dynamic . . . and combined with the performance of computing that allows for artificial intelligence to be applied, is going to bring all kinds of insights and additives to enhance everyday life."[15]
>
> —Brian Krzanich, Intel chief executive officer

The scale and variety of data are what make AI so powerful. As Intel chief executive officer (CEO) Brian Krzanich says, "Data is going to be dynamic . . . and combined with the performance of computing that allows for artificial intelligence to be applied, is going to bring all kinds of insights and additives to enhance everyday life."[15]

Turning Data into Intelligence

All the data AI collects from devices, websites, or human-labeled examples becomes useful only when the system can learn from it. This is where machine learning comes in. Machine learning is the main way modern AI learns from data, allowing systems to identify patterns and make predictions instead of following fixed, hand-coded rules.

Rama Ramakrishnan, an AI expert at MIT Sloan, explains the reasoning behind machine learning: "The basic idea of machine learning is, it's a lot easier to collect data than to collect understanding."[16] In other words, instead of trying to teach AI the meaning behind the data, it is provided with examples it can learn from.

> "The basic idea of machine learning is, it's a lot easier to collect data than to collect understanding."[16]
>
> —Rama Ramakrishnan, AI expert at MIT Sloan

Once a model is set up, it is trained on examples that represent the task it needs to perform. For instance, a system designed to detect spam emails is shown thousands of labeled messages. It analyzes the samples, learns the characteristic patterns associated with spam and non-spam emails, and updates its filtering process to reduce errors as it reviews additional messages. After the initial training, the AI can apply what it has learned to new, unseen emails, a step

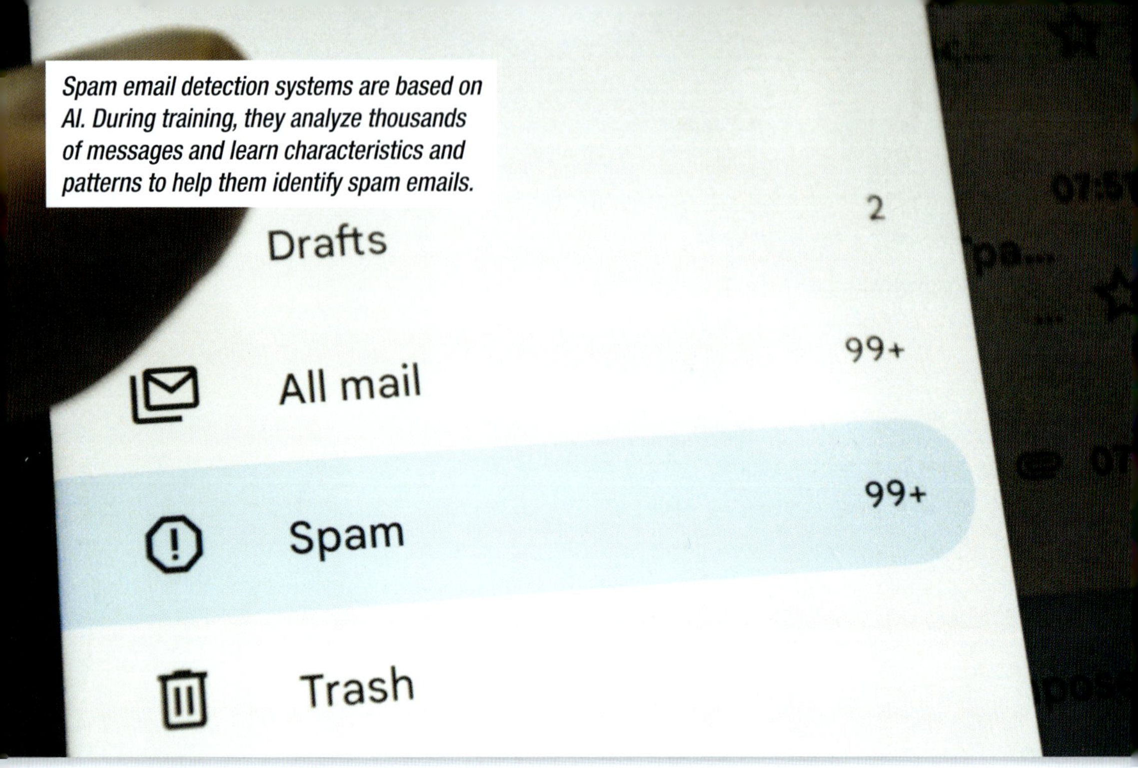

Spam email detection systems are based on AI. During training, they analyze thousands of messages and learn characteristics and patterns to help them identify spam emails.

known as inference. During inference, the AI makes predictions of whether a new email is likely to be spam based on the patterns it learned during training.

Today, machine learning is the backbone of most AI applications. For example, LLMs, which power generative AI tools like ChatGPT, are built using machine learning techniques; they are trained on massive datasets and then fine-tuned to generate new content. Machine learning allows AI to adapt to new information, making systems more flexible and able to handle larger tasks than early traditional approaches could. In this way, machine learning turns raw data into usable intelligence, allowing AI systems to perform complex tasks with increasing accuracy.

Types of Machine Learning

Machine learning comes in several forms, each suited to different kinds of problems. The three main types are supervised learning, unsupervised learning, and reinforcement learning.

Supervised learning is the most common. In this type, AI learns from labeled examples—data in which the correct answers are already known. For example, to teach a model to distinguish alligators from crocodiles, engineers provide thousands of images labeled "alligator" or "crocodile." The AI identifies patterns that separate one from the other and can eventually classify new images it has never seen. Supervised learning is used in common applications like email spam filters, image recognition, and speech-to-text systems.

Generative AI often uses a self-supervised variant of this approach, in which the "labels" are derived from the data itself, such as predicting the next word in a sentence. For example, when training a language model, the system is given part of a sentence and asked to predict a missing or next word. Because the original text already contains the correct word, the AI can compare its prediction to the real answer and adjust its calculations. By repeating this process across millions or billions of sentences, the AI gradually learns patterns in language, including grammar, meaning, and context.

Unsupervised learning works with unlabeled data, meaning the AI is not told the "right answers." It looks for hidden patterns, relationships, or groupings in the data. For example, an AI might analyze customer purchasing behavior and identify groups of shoppers with similar habits. These insights help companies

Deep Learning in Action: Studying Fruit Fly Development

Scientists often study fruit flies because their development happens quickly and follows clear patterns that are useful for understanding biology. In a recent study, researchers used deep learning to track how thousands of individual cells move and change during the earliest stage of a fruit fly's growth.

Using high-resolution videos of fruit fly embryos, the deep learning system learned how cells normally behave—where they move, when they divide, and how they interact with neighboring cells. Once trained, the model could predict, minute by minute, how each cell would fold, shift, or rearrange as the embryo developed. The system was accurate most of the time, even when tested on new videos it had never seen before.

This research shows how deep learning can help scientists study complex biological processes that are too fast or detailed for humans to analyze on their own. In the future, similar methods could be used to study how tissues and organs form in other animals or to spot early signs of diseases by identifying unusual cell behavior.

make marketing decisions, detect unusual sales activity, or discover trends.

Reinforcement learning takes a different approach. Instead of learning from examples, this type of AI learns through trial and error by interacting with an environment and receiving feedback in the form of rewards or penalties. As reinforcement learning researchers Richard S. Sutton and Andrew G. Barto explain in their book *Reinforcement Learning: An Introduction*, "Reinforcement learning is learning what to do—how to map situations to actions—so as to maximize a numerical reward signal."[17] In simpler terms, the AI tries different actions in a situation and observes the results. Actions that produce better outcomes are rewarded, while actions that produce worse outcomes are discouraged. Over time, the AI learns which actions are likely to lead to the highest rewards in similar situations. For example, a system learning to play a video game might try moving in different directions. Each move that earns points is reinforced, so the AI gradually discovers the strategy that maximizes its score. Generative AI can also use this approach when it is fine-tuned with human feedback, learning to produce responses that better match human preferences.

Deep Learning

Deep learning is a type of machine learning that helps AI solve very complex problems. It is especially useful when AI works with large amounts of data, like photos, videos, spoken language, or written text. Deep learning uses neural networks—computer models inspired by the human brain—that have many layers. Each layer looks at the data in a slightly different way, building understanding step-by-step.

For example, when analyzing an image, a neural network first notices simple features, like lines or edges. Then it combines these features to detect shapes, such as eyes, wheels, or letters. Finally, it connects the shapes to identify the full object, face, or word. Each layer adds more detail, helping the AI learn patterns from simple to complex.

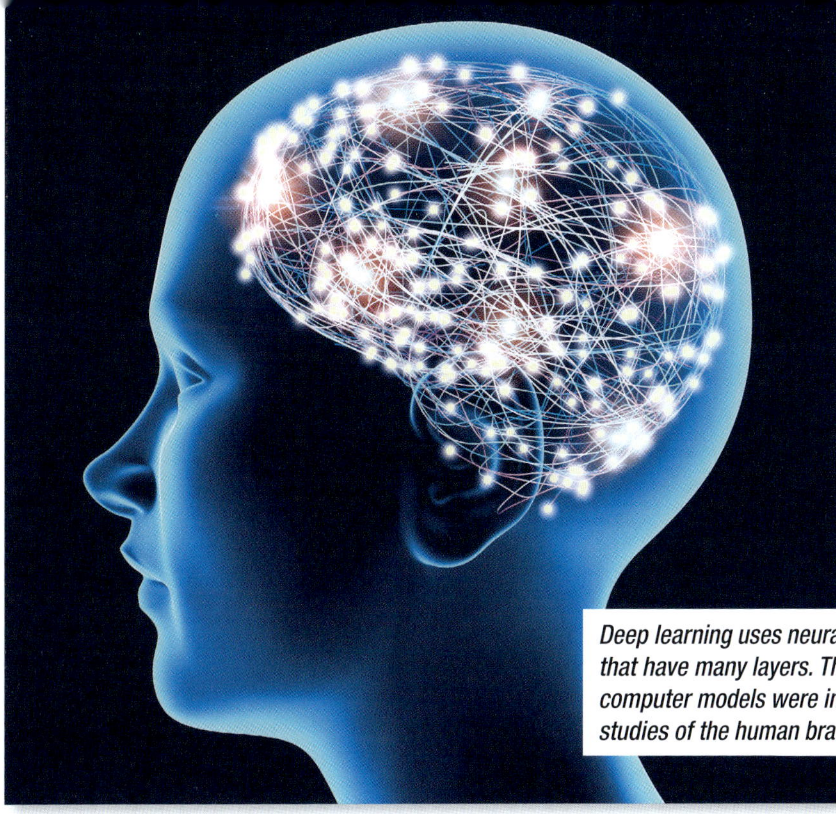

Deep learning uses neural networks that have many layers. These computer models were inspired by studies of the human brain.

This layered approach allows AI to do things that simpler machine learning systems cannot. Neural networks are used in facial recognition, voice assistants, speech-to-text tools, image search, and language translation. Large language models like ChatGPT also rely on neural networks to process and generate human language.

Training deep learning systems requires a lot of examples and powerful computers. During training, the AI makes predictions, checks whether they are correct, and adjusts its internal calculations to improve. Over time, it gets better at recognizing patterns and making accurate predictions. This process is what "learning" means for AI—it is based entirely on data and calculations, not understanding like a human.

How AI Learns Faster

Training deep learning systems needs not only large amounts of data but also the right hardware. Computers have processors, which follow instructions and perform calculations. The main

How Does AI Know If a Song Recommendation Worked?

When an AI system in a digital audio platform like Spotify recommends a song, it is making a prediction about what a listener might enjoy. To check how accurate that prediction was, the system observes what happens next.

The AI tracks behavioral signals such as whether the song is played, skipped after a few seconds, listened to all the way through, saved, or added to a playlist. These actions serve as feedback. Listening longer or saving a song suggests the recommendation was a good match, while skipping it quickly suggests it was not.

The system compares its original prediction with these outcomes. If listeners with similar habits respond positively, the recommendation is treated as successful. If they do not, the system adjusts its internal filters. Over time, these adjustments help the AI make better recommendations. This process relies entirely on patterns and feedback, not understanding, emotions, or personal judgment.

processor in most computers is the CPU, or central processing unit. CPUs are flexible and can handle a few complicated tasks at a time, like running programs or opening apps.

Deep learning requires millions of calculations at once, which is too much for a CPU to handle efficiently. That is why AI researchers use GPUs, or graphics processing units. GPUs were originally made to create images and video, which involves doing many calculations at the same time. This ability to handle thousands of calculations at once makes GPUs ideal for training large neural networks quickly.

Modern AI systems, including transformers, rely heavily on GPUs. Transformers can process large amounts of data at once, and GPUs make it possible to train them efficiently. This allows AI to detect complex patterns in text, images, and speech much faster than before. Without GPUs, training many of today's AI breakthroughs—from large language models to image recognition tools—would take weeks or months instead of days or hours.

Challenges in AI Learning

Learning is not always easy for AI. Even though these systems can find patterns in data, they face challenges that can make their predictions less accurate or reliable. Understanding these issues is important for anyone working with AI.

One common problem is bias in data. If the training data is incomplete or not varied enough, the AI may produce skewed results. For example, a language translation system trained mostly on formal text might struggle to understand casual speech or slang. Developers address this by using datasets that represent many different examples and by creating extra variations of the data, called data augmentation, to make it more balanced.

Errors are another concern. AI can make mistakes when it confronts situations it was not trained to handle. For instance, a self-driving car might misinterpret unusual road conditions it has never encountered. Developers reduce such errors by testing AI on separate datasets, adjusting the algorithms, and fine-tuning the models before using them in real-world situations.

AI can also overfit, meaning it becomes too focused on the training data and struggles with new examples. For example, an AI trained to recognize cats might do well with pictures taken from one angle or in bright light but fail with different perspectives or lighting. Developers prevent overfitting by carefully testing models on new examples and adjusting them so that they generalize better.

Finally, many AI systems require enormous amounts of data and computing power. Without enough data, it becomes difficult for AI to learn complex patterns. Researchers have warned that this reliance on ever-larger datasets may soon become a limiting factor. For example, in a 2024 study published by Epoch AI, a nonprofit research institute, the authors found that if current trends continue, "[large language] models will utilize the full supply of public human text data at some point between 2026 and 2032,"[18] making data availability a constraint on further progress.

This does not mean AI will stop improving, but it does present a challenge. Instead of simply using more data, researchers will need to focus on new approaches, such as training models more efficiently, reusing existing knowledge, and generating high-quality synthetic data. These strategies aim to help AI continue learning even when access to human-generated data becomes more limited.

How AI Predicts Words

When a person starts typing a text message and the keyboard suggests the next word, or when a chatbot completes a question with a full response, AI is quietly at work predicting language. These systems have studied millions of sentences to learn patterns in word sequences, context, and grammar, allowing them to generate text that fits naturally with what comes before.

AI that produces text—often called generative AI—relies on methods like natural language processing (NLP) and deep learning to identify these patterns. From predictive typing and virtual assistants to automated email responses and story generation, AI applies what it has learned to produce coherent language. Observing how words typically appear together allows these tools to complete phrases, answer questions, and even craft original sentences.

Understanding Natural Language Processing

NLP is the branch of AI that allows computers to understand and work with human language. It acts as a bridge between human communication and machine analysis, turning words and sentences into patterns that a computer can process. Every time someone uses predictive text on a phone, asks a virtual assistant a question, or chats with a help bot online, NLP is working behind the scenes.

NLP is a field of AI, not a type of machine learning or deep learning on its own. Instead, NLP systems often use machine learning to detect patterns in text and make predictions, such as classifying emails as spam or identifying the sentiment of a review. More advanced NLP, including LLMs like ChatGPT, relies on deep learning, which allows the AI to look at entire passages of text, recognize patterns, and generate responses that make sense and fit the passage.

NLP has two main functions: understanding and generation. Natural language understanding helps AI figure out what text means. It can identify the topic of a sentence, detect sentiment, or recognize key details such as names, dates, and locations. For example, natural language understanding allows a chatbot to know that the question "Where is the nearest library?" is about finding a location, not recommending a book.

Natural language generation allows AI to create new text based on what it has learned. Using natural language generation, AI can respond in a conversation, summarize long articles, complete a sentence a user started typing, or even write an original story. For instance, when a user types a few words in a messaging app and sees suggested completions, natural language generation is predicting the words that best fit the context.

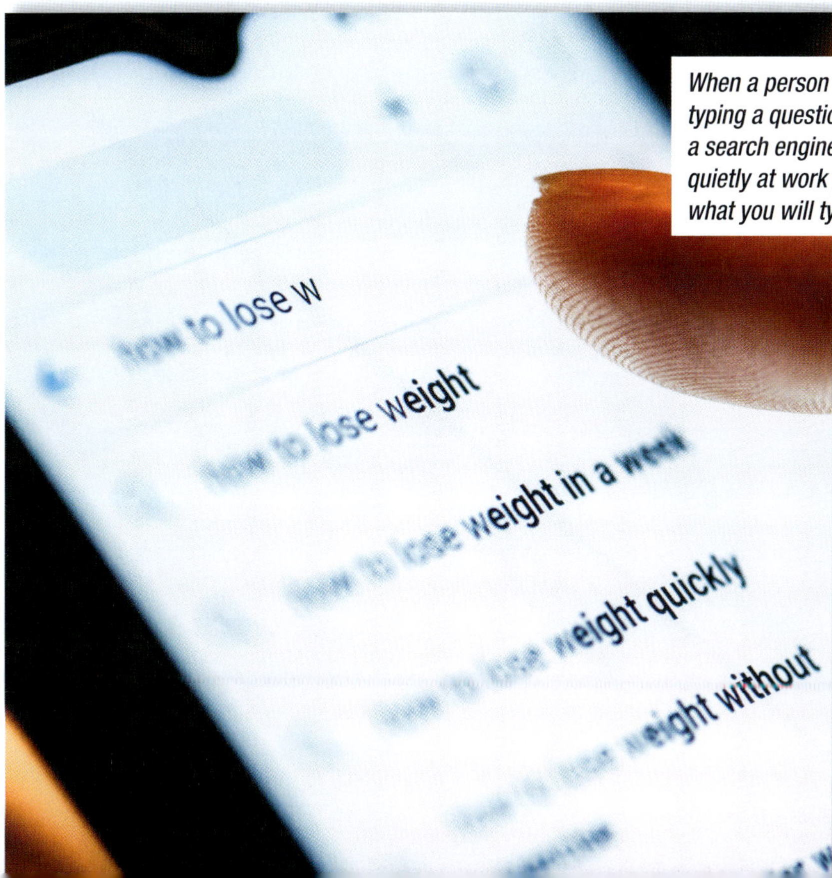

When a person starts typing a question into a search engine, AI is quietly at work predicting what you will type next.

To perform these tasks, NLP breaks text into smaller parts, studies grammar and sentence structure, and looks at how words relate to each other. Machine learning and deep learning allow the AI to detect patterns in language and apply them to new text, generating predictions and responses that are coherent and relevant.

NLP is also used in important real-world applications, such as health care. For example, researchers have applied NLP to electronic health records to automatically find patients who have had complex surgeries, like abdominal aortic aneurysm repairs. By analyzing doctors' notes, NLP can identify which patients had surgery, determine whether it was a first-time or repeat procedure, and organize this information for national databases. In a 2025 study published in *Scientific Reports*, some NLP models correctly classified nearly all cases while processing thousands of records quickly. This shows how NLP can go beyond simple text prediction to extract detailed, useful information from large amounts of unstructured text.

Language Models

Language models are the tools that allow AI to apply natural language processing on a large scale. They are trained on massive collections of text—everything from books and news articles to social media posts—so they can recognize patterns in word sequences, grammar, and meaning. By analyzing these examples, language models learn which words and phrases are likely to appear together, how sentences are structured, and how context shapes meaning.

For instance, a language model might learn that the word *bat* is often followed by *flies* in one context (talking about the animal) but *swings* in another (talking about baseball). By studying millions of sentences, the model learns to predict words and phrases that make sense in a given context. This predictive ability allows it to generate coherent text, complete sentences, or answer questions.

However, this process does not guarantee that a model treats all information consistently. In fact, models can answer two related questions differently, not because they are "confused" in the human

Language models are trained on massive collections of text from books, magazines and even social media posts, so they can recognize patterns in word sequences and meaning.

sense, but because different parts of the model are being used for each question. One part might store knowledge about a fact, while another part evaluates whether a statement is true. Josh Batson, a researcher at AI system developer Anthropic, explains, "It's much more like: 'Why does page five of a book say that the best food is pizza and page 17 says the best food is pasta? What does the book really think?' And you're like: 'It's a book!'"[19] In other words, language models can appear contradictory because they rely on multiple internal mechanisms, each of which can produce different outputs depending on the way a question is asked.

Text Prediction

Text prediction is how AI guesses the next word or phrase in a sentence, continues a conversation, or answers a question. For example, when someone types "I went to the . . ." on a phone, predictive text might suggest words like *store*, *park*, or *movies*.

The AI makes these predictions by looking at patterns it has learned from millions of examples, including books, articles, and

What Happens When Grammar or Punctuation Is Poor?

AI language models are trained on vast amounts of real-world text, much of which includes misspellings, informal grammar, incomplete sentences, and inconsistent punctuation. Because of this exposure, AI learns that language is often messy and that meaning does not depend on perfect writing.

When someone types with poor grammar or punctuation, the AI looks beyond surface errors and focuses on patterns of word use and context. It analyzes which words tend to appear together and how similar phrases are used in other situations. For example, even if a sentence is missing punctuation or uses the wrong verb tense, surrounding words often provide enough clues for the AI to infer intent.

Transformers help with this process by allowing the model to consider the full sentence or conversation rather than interpreting each word in isolation. This broader context makes it easier to distinguish mistakes from meaning. While AI can still misinterpret unclear or highly ambiguous text, its ability to recognize patterns allows it to respond accurately in many imperfect, everyday writing situations.

emails. It also considers the words the user is currently typing to improve accuracy. For instance, if a conversation mentions a roller-skating rink, the AI may suggest "roller-skating rink" in related sentences during that session. However, this knowledge is temporary unless the AI is specifically trained or personalized to remember it.

Text prediction works through NLP and machine learning. NLP allows the AI to break down text, understand grammar and context, and see how words relate to each other. Machine learning, especially deep learning, lets the AI recognize patterns in large amounts of text and calculate which word is most likely to come next. Together, these technologies let AI generate text that makes sense in the sentence. For example, if the AI sees "I like to eat pizza and . . ." it can predict that the next word might be "pasta" or "salad" rather than something unrelated like "car" or "phone."

By repeating this process for each word, AI can complete sentences, generate paragraphs, or respond to questions. It may feel almost magical because it seems to anticipate the user's thoughts, even though it only works with patterns and probabilities, not actual understanding.

Text prediction powers many everyday tools beyond phone keyboards. Chatbots, virtual assistants, email composition tools, and educational writing aids all rely on this technology. In some systems, AI can even improve during a session or through personalization, making suggestions feel more tailored to the individual user.

Transformers in NLP

Text prediction works well for completing single words or short phrases, but language often stretches across multiple sentences or even paragraphs. To generate longer, coherent text or follow a topic over time, AI needs a way to keep track of context across many words. This is where transformers come in.

Transformers allow models to process many words at once instead of one at a time. By looking at the full passage, they can see connections between words, sentences, and ideas. This makes it possible for AI to produce longer, more coherent responses that stay on topic.

Jensen Huang, founder and CEO of Nvidia, describes the impact of transformers on AI: "Transformers made self-supervised learning possible, and AI jumped to warp speed."[20] This shows how transformers helped AI learn faster and handle more complex language tasks.

> "Transformers made self-supervised learning possible, and AI jumped to warp speed."[20]
>
> —Jensen Huang, founder and chief executive officer of Nvidia

Transformers are sometimes compared to human working memory, which lets people hold information temporarily—like remembering the start of a sentence while reading the end. Transformers do not think or understand meaning, but they use attention to track important words and ideas across a passage. For example, if a person or topic is mentioned early, the AI can keep referring to it later instead of treating each sentence as separate. By tracking context across longer stretches of text, transformers let AI generate responses that stay connected, clear, and focused, even in long paragraphs or extended conversations.

Task-Specific Training

By the late 2010s, advances in transformers allowed AI to process long passages of text, but these models were still general-purpose—they could generate language, but they were not specialized for particular tasks.

Task-specific training is a form of machine learning that refines a general language model for a precise purpose. This is accomplished by focusing the AI on one task within a dataset. For instance, a translation tool is trained on millions of paired sentences in two languages so it can learn how meaning carries across words and grammar. A writing assistant can be trained on user drafts paired with professionally edited versions, showing it how to improve clarity, tone, and sentence flow. Even medical chatbots can be trained on thousands of select patient–doctor interactions, helping them generate accurate answers to questions within a specified area of medicine.

By emphasizing examples drawn from a specific domain, task-specific training steers the model's output toward the conventions that are most relevant to users' needs. This might help mechanical engineers adopt résumé formats that are successful in their field or aid translators in recognizing subtleties when shifting from Arabic to American political discourse.

Human feedback often guides this training. Reviewers flag incorrect or unclear responses or indicate which outputs are most helpful. The AI uses this information to adjust its patterns, producing better results over time. Task-specific training does not give the AI understanding or intent, but it makes its word choices more accurate, consistent, and reliable.

In short, task-specific training is machine learning applied to words, transforming a broad language model into a specialized tool capable of performing real-world language tasks effectively.

Applications in Everyday Tools

Once language models have been refined through task-specific training, they begin to appear in tools people use every day. These

Chatbots such as ChatGPT use language models trained for conversation to answer questions, explain topics, and generate written responses that follow context.

applications apply learned language patterns in focused ways, turning abstract training into practical features that support communication, productivity, and access to information.

Chatbots such as ChatGPT are one example. These systems use language models trained for conversation to answer questions, explain topics, and generate written responses that follow context across multiple turns. Virtual assistants rely on similar language capabilities but are further trained to recognize commands, retrieve information, and interact with other software, such as calendars, reminders, or smart home devices.

Grammar and writing assistants focus on sentence structure, clarity, and tone. Through task-specific training, these tools learn to identify patterns associated with common errors or awkward phrasing and suggest revisions based on how similar sentences are typically written. As a result, they help users improve readability and consistency across emails, essays, and other written materials.

Challenges in Language Understanding

Even though AI language tools are useful in many everyday situations, they still face important challenges. These challenges come from the way AI learns language—by finding patterns in large amounts of text rather than understanding meaning the way people do. As a result, AI can struggle when language relies on tone, cultural knowledge, or shared human experience.

One common difficulty is interpreting sarcasm, humor, and idioms. Emily Bender, a computational linguist and expert in the linguistic foundations of AI, explains why AI struggles with these forms of language: "[LLMs] see 'amazing' paired with negative events—but they don't grasp the speaker's commitment to irony, or their expectation that the listener will recognize it."[21]

> **"[LLMs] see 'amazing' paired with negative events—but they don't grasp the speaker's commitment to irony, or their expectation that the listener will recognize it."[21]**
>
> —Emily Bender, computational linguist and expert in the linguistic foundations of AI

For example, phrases like "nice job" can mean very different things depending on context and tone. Humans usually understand these differences instantly, but AI must guess based on how similar phrases appeared in its training data. This shows that while AI can detect patterns, it cannot understand intention or tone, which makes sarcasm and irony especially challenging.

Another challenge involves bias in training data. Because AI learns from text written by people, it can pick up harmful stereotypes or offensive ideas that appear often in those sources. Timnit Gebru, computer scientist and former co-lead of Google's Ethical

How Transformers Made Generative AI Possible

Generative AI refers to AI systems that can produce original text, images, or code in response to a prompt. ChatGPT is a well-known example of generative AI. Earlier AI systems could predict the next word in a sentence, but they struggled to generate long, meaningful responses. Their predictions were based on short sections of text, which caused them to lose track of topics, repeat information, or produce disconnected answers.

Transformers solved this problem by allowing AI models to consider much larger amounts of text at once. Instead of focusing only on the most recent words, a transformer-based system like ChatGPT can take into account what was written earlier in a passage, including important topics, names, and questions. This allows the AI to stay focused and respond in a more organized way.

Because transformers made this possible, developers were able to train large language models on vast amounts of written material, such as books, articles, and websites. These models can then generate complete explanations, conversations, and written responses. Generative AI still works by predicting words, but transformers allow those predictions to form clear, connected answers.

AI team, puts it simply: "If you train an AI on biased data, it will give you biased results."[22] This has led to cases in which AI systems repeat racist, misogynistic, or otherwise biased language. These outputs are not intentional, but they reflect the patterns present in the data the AI learned from.

To reduce these issues, developers use methods such as filtering training data, testing for biased outputs, and applying human feedback to improve responses. While these steps help, understanding complex and nuanced language remains one of the hardest problems in AI research.

CHAPTER FOUR

How AI Sees and Creates Images

A person can glance at a photo and immediately recognize a friend, a dog, or a familiar place. This feels effortless because the human brain is naturally specialized for visual recognition. For AI, the same task is much more complicated. When an AI looks at a picture of a cat, it does not see whiskers, fur, or eyes the way a human does. Instead, it receives a grid of numbers that represent light and color. From these numbers, the AI learns patterns that often appear together in images labeled as "cat," gradually allowing it to recognize similar images.

Computer vision is the field that makes this possible. It allows AI to find structure and meaning in raw pixels, gradually learning which visual patterns tend to appear together and what they usually represent. The same pattern-learning process also makes it possible for AI to generate new images that resemble photos, paintings, or drawings it has studied before. Through these methods, machines can recognize faces, identify objects, and even create original visuals, all by detecting and recombining patterns in image data rather than truly seeing or understanding the world the way humans do.

Understanding Computer Vision

Computer vision is the branch of AI that allows machines to interpret the world through images and videos. Unlike humans, AI does not have eyes or a brain to naturally make sense of what it sees. Instead, it looks at pictures as grids of tiny dots called pixels and learns to detect patterns, shapes, colors, and the relationships between objects. This lets AI identify what is in a photo or video without "seeing" it like a person does.

Images carry information differently than text. A computer must understand not just what objects are present, but where they are, what they are doing, and how they relate to other objects. This focus on spatial and visual relationships is what makes computer vision unique.

AI learns to do this using deep learning and neural networks, which are systems inspired by the human brain. These networks can be trained to recognize objects, faces, movements, and even subtle patterns that humans might overlook. Once trained, they can apply what they have learned to new images with impressive accuracy.

Computer vision is used in many real-world applications. Self-driving cars rely on it to detect other vehicles, pedestrians, and traffic signals. Social media apps use it to apply face filters or automatically tag people in photos. Hospitals use it to analyze X-rays and magnetic resonance imaging scans, helping doctors detect

When an AI looks at a picture of a cat, it does not see whiskers, fur, or eyes the way a human does. Instead, it receives a grid of numbers that represent light and color.

problems more quickly. Factories use it to check products for defects, and drones use it to monitor crops or survey disaster zones.

Computer vision can run on cloud servers, on local computers, or directly on devices like phones and robots. This flexibility allows AI to process images in real time, even in places without fast internet. Researchers are now working on systems that can better understand context, combine visual data with other information, and improve their own learning with less human guidance. These advances could make computer vision smarter, faster, and more reliable in everyday situations.

How Computer Vision Recognizes Images

Computer vision relies on deep learning models, especially convolutional neural networks, to understand images. These networks process an image in layers. Early layers detect simple features, such as shapes, colors, or basic textures. Later layers

Convolutional neural networks process images in layers, then combine them to create images, like this AI image of a dog.

> "I was shocked that this was so easy to do—to make the models think the wrong thing."[23]
>
> —Andrew Ilyas, computer scientist

combine these details to identify patterns, textures, and ultimately whole objects like cars, dogs, or buildings. In this way, the AI gradually builds a more complete representation of what it "sees."

Computer vision can do more than just recognize one object at a time. For example, when looking at a video of a park scene, AI can identify each object—a tree, a dog, or a bench—in the image and determine their spatial relationships to each other. It can even separate objects that are close together, distinguishing the dog from a cat sitting nearby. If the dog is running, the AI can follow it from frame to frame, keeping track of its movements. By combining these abilities, the AI can understand the whole scene and how the different objects interact with each other.

Even with these sophisticated techniques, computer vision models can make surprising mistakes. Despite their strong performance, neural networks sometimes misinterpret images in ways that seem baffling to humans. "I was shocked that this was so easy to do—to make the models think the wrong thing,"[23] says computer scientist Andrew Ilyas, a member of a student research team at MIT. In 2017 his group tricked a neural network into labeling a picture of a cat as guacamole by adding just a tiny amount of pixel "noise" to the image. The change was almost invisible to people, but it completely confused the AI.

This example shows that while deep learning gives computers powerful image-recognition abilities, these systems are still working with patterns and probabilities, not true understanding. They can recognize complex objects and scenes, but they remain sensitive to subtle changes in the data, highlighting the unique challenges of teaching AI to "see" like humans do.

Generating Images with AI

AI can do more than recognize images; it can create entirely new ones. In text-to-image generation, a user types a description, and the AI produces a matching picture. For example, if someone types “a penguin painting a rainbow in a snowy landscape,” the AI can combine those ideas into a realistic or artistic image. To do this, it must learn not just what individual objects look like, but also how shapes, colors, textures, and perspectives fit together.

Early AI image generation relied on generative adversarial networks (GANs), in which one network creates images and the other checks whether they look real or fake. If the second network judges that the image appears fake, then it will provide feedback that informs how future images are produced. This process improved results over time, but it still struggled with complex or highly detailed images. Some modern systems use more advanced deep learning methods rather than GANs. These systems find patterns in colors, shapes, lighting, and how objects are organized across vast image datasets.

One approach, diffusion models, starts with random “noise”—like the static on an old television screen—and gradually refines it into a clear image, guided by patterns learned from millions of examples. With each step, the model adjusts the pixels in the noise so that shapes, colors, lighting, and details slowly emerge. By the end of the process, these refinements combine to form an image that closely resembles a realistic scene, though not one that necessarily matches any specific training example. As Hany Farid, a computer science professor at the University of California, Berkeley, explains, “All of the [training] images are relevant to [the model]. Everything it's been trained on, the whole body of it. And that's what makes it so amazing. It's not cobbling together an image from a bunch of images it's seen before—that'd be really boring.”[24]

Variational autoencoders, or VAEs, simplify an image into a compressed version that captures its essential features. The system can then rebuild, or decompress, the image when needed, or generate a new image that is similar but not identical. For example, a VAE trained on cars might produce an automobile with a familiar shape but a different color or body style.

Image-based transformers treat parts of an image like words in a sentence, analyzing how the pieces fit together to produce

How Computer Vision Works in *Pokémon GO*

In *Pokémon GO*, players explore the real world through their phones, and the game overlays virtual creatures onto the camera's view. The AI doesn't "see" the world like a human; instead, it analyzes the camera feed as digital data, identifying surfaces, shapes, and movement. This allows the game to place Pokémon on the ground or attach them realistically to objects in the environment.

The AI also tracks the player's location using the Global Positioning System and detects movement to decide when and where Pokémon should appear. Using these inputs, it determines how creatures should interact with the surroundings, such as appearing on a sidewalk rather than floating in midair. When players try to catch a Pokémon, the system calculates depth, distance, and perspective to make throws feel accurate. The game demonstrates the core of computer vision: sensing the environment, interpreting data, and guiding actions—showing how AI bridges digital and physical worlds.

detailed and coherent visuals. For example, if the model sees part of an image that looks like a face, it can predict where eyes, a nose, and a mouth should appear in relation to one another based on learned patterns. This allows transformers to maintain consistency across the entire image, helping them produce detailed scenes in which objects fit together and maintain appropriate spatial relationships and lighting.

Everyday Applications of Computer Vision

Computer vision powers many everyday technologies, turning ordinary devices into smarter tools that can see and respond to the visual world. One of the most familiar ways computer vision appears is in smartphone cameras. Cameras can detect faces, adjust focus, and suggest filters to improve photos. Some phones can even notice smiles, blinking, or multiple people in a frame, helping users take sharper, better-composed shots.

Social media apps build on this by tracking facial features in real time. Users can try virtual makeup, animal ears, or other interactive filters that move with them. Augmented reality effects can place digital objects in a room, like furniture or decorations, and align them with walls, floors, or tables.

Computer vision powers social media apps that track facial features in real time. Users can try virtual makeup or other interactive filters that move with them.

Computer vision also powers creative tools. Apps that generate artwork or memes can identify objects, colors, and patterns in images and then combine them in new ways to make original visuals. Video-editing apps can remove backgrounds, track moving objects, or add effects automatically. These tools make it possible for anyone to create polished videos or digital art without special training.

Gaming and entertainment benefit from computer vision as well. Games that incorporate motion-tracking cameras can read players' gestures or body movements to control the action. Augmented reality games overlay digital characters or objects into real-world spaces, allowing players to interact with them—walking around a virtual dragon or placing a puzzle on a table.

Limitations and Mistakes in AI Vision

Although computer vision has made remarkable progress, it remains far from perfect. AI systems do not actually understand images in the way humans do. Instead, they analyze patterns in pixels and match them to examples seen during training. Because of this, even advanced models can misidentify objects, confuse similar shapes, or fail when lighting, angles, or backgrounds change in unexpected ways. In generated images, these limits often appear as distorted hands, mismatched facial features, or details that look almost right but subtly wrong.

One major limitation is that AI struggles to adapt to new environments it has not seen before. Russian American computer scientist Alexei Efros highlights this gap by comparing AI to living beings in the real world. "If a field mouse ends up in a house, it'll do fine," he explains. "It was born in a field, has never been in a house before, and yet it will find and eat all of your supplies. It adapts very quickly, learns and adjusts to the new environment."[25] Current computer vision systems do not have this flexibility. Efros points out that a self-driving car trained in California may fail when tested in Minnesota because of snow—an environment it was never trained on—causing the system to become confused. This shows how strongly visual AI depends on past data rather than true understanding.

Another concern is how this technology may change art culture more broadly. Since image generation can happen instantly from a short prompt, some fear that automation may replace

How DALL-E Turns Words into Images

DALL-E is an AI tool that can create pictures from written descriptions. When a user types a prompt, like "a cat reading a book on the moon," the AI first breaks the text into smaller concepts, understanding objects, actions, and styles.

Next, the AI draws on patterns it learned from millions of images and captions. Diffusion models, which power DALL-E, start with random "noise"—a messy cloud of pixels—and gradually refine it into a coherent image. The model adds shapes, colors, textures, and details step-by-step, checking each part against what it has learned until the final picture matches the prompt.

DALL-E does not copy existing images; it generates something new based on learned patterns. This allows it to combine ideas in ways humans might not imagine. While the AI can produce impressive and realistic visuals, its results can sometimes include small mistakes, like awkward proportions or unusual details, because it doesn't truly "see" the world—it only interprets patterns in data.

"I foresee with trepidation a future where the use of Generative AI becomes the norm. More people putting in prompts to have detailed 'artworks' appear, than artists working with pencils, pens and brushes."[26]

—David Robertson, comics creator

human effort rather than support it. "I foresee with trepidation a future where the use of Generative AI becomes the norm. More people putting in prompts to have detailed 'artworks' appear, than artists working with pencils, pens and brushes,"[26] says David Robertson of Fred Egg Comics.

Concerns about creativity also extend to how AI imitates existing styles rather than developing new ones. Generative models do not invent ideas from experience or intention. They recombine patterns from millions of training images according to probability. Comics creator Chloe Starling describes the difference between how humans and AI models approach art: "When people are inspired by previous work, they build upon it and develop it in a human, organic way that moves the art form forward. But [generative] AI simply uses algorithms and code to imitate art, spitting out cheap imitations made with no soul."[27]

Beyond creativity, there are also serious risks related to accuracy and trust. Generative models can produce images that look realistic but show events that never happened. Because the system has no concept of truth, it cannot tell the difference between a real scene and a believable fake. This makes computer vision vulnerable to misuse in propaganda, scams, and misinformation. Hany Farid warns, "You can see why this is in some ways a much more powerful technology from the point of view of creativity, but also manipulation."[28]

Other challenges remain as well. Visual AI can reflect biases in its training data, perform poorly on underrepresented groups, and struggle in unfamiliar situations. It can appear confident even when it is wrong, and it cannot explain its decisions in human terms. These limitations remind us that while AI vision is powerful, it is still a tool built from data and algorithms—not a system that truly sees, understands, or adapts to the world on its own.

CHAPTER FIVE

The Full Potential of AI

It is Friday night, and Ashley is babysitting five-year-old Ben. He asks for a bedtime story, but they have already read every book on the shelf. Ashley pulls out her iPad and asks what kind of adventure he wants. Ben grins and says, "A kangaroo who becomes friends with a lizard, and they go looking for ice cream."

Ashley types the idea into an AI tool, and almost instantly, a full story comes to life. The AI builds the plot, writes the dialogue, and creates vivid illustrations of the characters in action—all at the same time. By linking reading, analyzing, and creating in a single process, AI transforms a simple idea into a rich, imaginative experience.

AI That Combines Skills

Many modern AI systems go beyond performing one task; they combine multiple abilities they have learned. This is often called multi-skill AI. For example, an AI in a self-driving car can simultaneously read traffic signs, track nearby vehicles and pedestrians, and plan the best route. Each skill is trained separately, and then the AI tool learns to link them so it can perform all of them together. By combining skills, AI can handle tasks that no single ability could manage alone.

Many multi-skill AI systems are also multimodal, meaning they can process different types of information at once, such as text, images, video, and audio. Multimodal AI is what allows tools like smart assistants or advanced robots to take a picture, interpret spoken instructions, and generate responses that include words, images, or sounds. Some well-known tools, such as DALL-E, are multimodal in a limited way: They take text as

> “Humans are already multimodal. Our ability to build embodied AI models that can see, hear, understand language and handle other sensory tasks is crucial for development of far more sophisticated technology.”[29]
>
> —Yoon Kim, assistant professor of electrical engineering and computer science at the Massachusetts Institute of Technology

input and generate images as output, linking language understanding with visual creation, even though they do not process video or audio.

Yoon Kim, an assistant professor in the Electrical Engineering and Computer Science Department at MIT says, “Humans are already multimodal. Our ability to build embodied AI models that can see, hear, understand language and handle other sensory tasks is crucial for development of far more sophisticated technology.”[29] This highlights how AI aims to mimic the way humans combine multiple senses and skills, enabling machines to interpret and act on complex information in a coordinated way.

These systems work by first learning patterns from large datasets for each skill or input type. When they are used, the AI combines those patterns to produce coordinated outputs—such

Many modern AI systems use multi-skill AI. For example, AI in a self-driving car can simultaneously read traffic signs, track vehicles and pedestrians, and plan the best route.

as decisions, alerts, or images that fit together logically. Multi-skill and multimodal AI continue to develop, and future systems are likely to handle more types of information at once, allowing AI to take on increasingly complex tasks.

Linking Different Abilities

To combine different abilities, many AI systems rely on more than one model working together. A model is a trained system that specializes in a specific task, such as understanding language or recognizing objects in images. Instead of one single model doing everything, complex AI systems often connect several models, each trained for a different type of data or skill.

All inputs—such as text, images, or audio—are first converted into numbers. This allows the models to process very different types of information in a similar way. One model may analyze an image to detect objects, shapes, or actions, while another focuses on understanding words and sentences. Each model produces results based on patterns it has learned during training.

A coordinating system then links these results together. For example, an image model might identify a "bicycle" and a "person," while a language model uses that information to describe the scene or answer a question about it. The models share information, even though they were trained separately, allowing the AI to produce responses that combine visual and language understanding.

This structure makes multi-skill AI possible. Instead of relying on a single ability, the system connects specialized models to generate outputs that appear unified. Sparkli, a learning platform for children aged five to twelve, demonstrates this approach in practice. Its AI combines visuals, voice, and interactive simulations to create personalized learning experiences, adapting to each child's interests and choices. As CEO Lax Poojary explains, "Our goal is to build agency in the next generation. Children learn by exploring, making choices, asking questions, and discovering what inspires them,"[30] guiding experiences that naturally integrate multiple skills and senses.

The AI is not thinking or understanding like a human. Instead, it links patterns across multiple models to produce outputs that fit together logically, whether for answering questions, describing scenes, or guiding a child through an interactive learning adventure.

Autonomous AI Agents

As AI systems gain the ability to link different skills, a new layer is needed to manage how those skills work together. This is where autonomous AI agents come in. An AI agent is a type of software that can act on its own to achieve a goal.

The agent starts with an overall mission and breaks it into smaller tasks. Each task might use a different skill, like understanding language, analyzing images, searching databases, or making predictions. The agent picks the right skill for each step. For example, it might read a message, check information in a database, and interpret an image. Then it uses all that information together to decide what action to take next. If new information appears, it updates the plan instead of starting over. This allows the agent to keep working continuously.

These agents are already being used in workplaces. As Helen Poitevin, senior research director at Gartner, explains, "Digital workplace leaders will proactively implement AI-based technologies such as virtual assistants or other NLP-based conversational agents and robots to support and augment employees' tasks and productivity."[31] She warns, however, that careful oversight is needed. Without proper rules and monitoring, these systems can produce frustrating or harmful experiences—for example, sending repeated unwanted notifications or making automated messages that feel intrusive.

> **"Every [AI] agent should have similar security protections as humans, to ensure agents don't turn into 'double agents' carrying unchecked risk."[32]**
>
> —Vasu Jakkal, corporate vice president of Microsoft Security

Because autonomous agents can act independently and affect real systems, safeguards are essential. "Every agent should have similar security protections as humans, to ensure agents don't turn into 'double agents' carrying unchecked risk,"[32] says Vasu

How a Robot Vacuum Works

A robot vacuum is more than a moving vacuum cleaner—it is an autonomous system controlled by an AI agent. The agent acts like a small digital brain, deciding where to go, what to clean, and how to adjust when obstacles appear. It starts with a goal: clean the entire floor. The agent breaks this goal into smaller steps, such as detecting walls, mapping rooms, and navigating around furniture.

Sensors feed information to the AI agent, including the location of obstacles, floor type, and stair edges. The agent processes this data and selects the right action for each step. For example, it may slow down near fragile objects or change direction to cover an area it missed. The AI agent continuously updates its plan based on new information, which allows the vacuum to operate efficiently without human guidance.

Advanced robot vacuums use mapping algorithms and memory to plan paths and avoid cleaning the same area twice. The AI agent also manages battery levels, automatically returning to the charging station when power is low. By combining perception, reasoning, and action, the robot vacuum demonstrates how AI agents coordinate multiple skills to achieve a goal in the real world.

Jakkal, corporate vice president of Microsoft Security. Without proper controls, an agent could make mistakes—like sharing private information with the wrong person or triggering an action at the wrong time. Defining what an agent can access, what actions it can take, and how its behavior is monitored helps prevent accidents and misuse.

Autonomous agents are especially useful in complex systems such as self-driving cars, robotics, and advanced digital assistants. In these situations, no single skill is enough. The agent must coordinate perception, reasoning, and action in real time. By managing multiple AI abilities safely and effectively, autonomous agents demonstrate how AI is moving beyond isolated tasks toward fully integrated intelligence.

Types of AI Agents

AI agents are systems designed to make decisions and take actions. Some agents follow simple rules, while others can plan, weigh options, or learn from experience. The type of agent used depends on how complex the task is and how much flexibility is needed.

The simplest kind is a reflex agent. These agents react immediately to what they sense, using fixed rules. They do not remember past events or adjust their behavior over time. For example, a thermostat turns on the heat when the temperature drops below a set point. If conditions change in an unexpected way, the agent cannot adapt.

A model-based reflex agent adds memory. It keeps an internal picture of its environment and updates that picture as new information appears. A robot vacuum works this way by remembering where it has already cleaned and avoiding obstacles as it moves.

Goal-based agents are designed to reach a specific outcome. Instead of acting right away, they consider different actions and choose the one that leads toward their goal. A navigation app that plans a route to a destination is a common example.

Utility-based agents also work toward goals, but they focus on choosing the best option rather than just any option. They

A robot vacuum keeps an internal picture of its environment and updates that picture as new information appears. In this way, it can avoid obstacles as it moves.

How ChatGPT Uses Different Types of AI Agents

ChatGPT is built using ideas from several types of AI agents, rather than just one. Each type helps it handle questions and tasks in a different way.

First, ChatGPT works like a learning agent. It was trained on large amounts of text, which allows it to learn patterns in language and improve how it responds. This training helps it answer many kinds of questions, even new ones it has not seen before.

ChatGPT also behaves like a goal-based agent. When it receives a prompt, it focuses on a clear goal, such as answering a question, explaining a topic, or helping with writing. Every response is shaped around completing that goal.

ChatGPT also uses ideas from utility-based agents. It chooses responses that are more helpful, clear, and safe over ones that are less useful or inaccurate. This helps it select the best possible answer from many options.

These aspects work together to provide results that match a user's query. The purpose is to provide reasonable responses and learn from feedback how to generate better output in the future.

compare possible outcomes using a utility score, which measures how helpful or desirable each option is. For example, a navigation system might choose a route that slightly increases travel time but saves fuel, avoids toll roads, and reduces traffic stress.

Finally, learning agents can change their behavior over time. They learn from experience by storing past actions and outcomes, then using that information to make better decisions in the future. For example, a music or video streaming service learns what users like by tracking what they watch, skip, or replay. Over time, the agent improves its recommendations, offering suggestions that better match each person's preferences.

Challenges and Limitations

Multi-skill or multimodal artificial intelligence can struggle when it tries to combine different abilities into a single response. For example, an AI tool that works with both text and images might describe a picture incorrectly or mix details from unrelated topics. It may also try to solve a problem using knowledge from one area that does not apply to another. These mistakes can make outputs confusing, inconsistent, or even illogical.

Computers cannot match humans in their ability to reason abstractly in real world situations. Humans still need to oversee computers to solve the most pressing problems.

AI agents, which are designed to complete tasks on their own by taking multiple steps, can face similar difficulties. They may misunderstand the order of tasks, repeat actions unnecessarily, or make decisions that do not match what a user wants. Because AI relies on patterns in its training data rather than complex reasoning like a human, small errors can grow as tasks become more complex. As AI pioneer Michael I. Jordan explains, "For the foreseeable future, computers will not be able to match humans in their ability to reason abstractly about real-world situations. We will need well-thought-out interactions of humans and computers to solve our most pressing problems."[33]

These challenges mean that multi-skill AI and agents cannot always combine their abilities perfectly. Users must check outputs carefully, give clear instructions, and recognize that results may sometimes be unexpected or wrong. While these systems can handle impressive and complex tasks, their integration of multiple skills is not flawless, and mistakes are a natural part of how they operate.

Big Picture

At its core, artificial intelligence shows what is possible when many abilities work together. Instead of relying on a single skill, AI can connect language, images, and reasoning to produce answers, explanations, and creative outputs that are richer and more flexible than earlier systems. This ability allows AI to support a wide range of tasks, from problem-solving to content creation.

Even so, AI does not think like a human. It follows patterns from data rather than understanding ideas or intentions, which means it can still make mistakes, give inconsistent results, or misunderstand what is being asked. Human judgment remains essential, especially when decisions matter.

Together, these strengths and limits point to a clear role for AI in the future. As Aparna Chennapragada, Microsoft's chief product officer for AI experiences, explains, "The future isn't about replacing humans. It's about amplifying them."[34] By bringing multiple abilities together, AI can extend human knowledge, creativity, and problem-solving, offering new ways to explore, learn, and create.

SOURCE NOTES

Introduction: Meet AI: The Technology Shaping Our World

1. Quoted in Bernard Marr & Co., "28 Best Quotes About Artificial Intelligence," 2026. https://bernardmarr.com.
2. Quoted in Vahid, "Data Is the New Oil: A Powerful Metaphor with a Catch," Medium, May 23, 2025. https://medium.com.
3. Quoted in TED, "The Era of Blind Faith in Big Data Must End," 2017. www.ted.com.
4. Quoted in Tanmay Das, "Understanding Generative AI: Capabilities, Applications, and Future Impact," Karmick Institute, May 7, 2025. www.karmickinstitute.com.
5. Quoted in Gams Matjaz, "Evaluating ChatGPT's Consciousness and Its Capability to Pass the Turing Test: A Comprehensive Analysis," *Journal of Computer and Communications*, March 2024. www.scirp.org.

Chapter One: The Evolution of AI

6. Quoted in Amber Jackson, "Alan Turing: A Strong Legacy That Powers Modern AI," *AI Magazine*, June 7, 2024. https://aimagazine.com.
7. John McCarthy, "Professor John McCarthy: What Is AI?," Professor John McCarthy personal website, 1955. http://jmc.stanford.edu.
8. Quoted in AI World Society, "This Week in the History of AI at AIWS.net," March 5, 2021. https://aiws.net.
9. Quoted in Lee Dembart, "Experts Argue Whether Computers Could Reason, and If They Should," *New York Times*, May 8, 1977. www.nytimes.com.
10. Judea Pearl, *The Book of Why: The New Science of Cause and Effect*, May 15, 2018. www.goodreads.com.
11. Quoted in Arc Institute, "AI Can Now Model and Design the Genetic Code for All Domains of Life with Evo 2," February 19, 2025. https://arcinstitute.org.
12. Quoted in Sara Brown, "Machine Learning, Explained," MIT Sloan School of Management, April 21, 2021. https://mitsloan.mit.edu.

13. Quoted in Rob Towes, "Transformers Revolutionized AI. What Will Replace Them?," *Forbes*, September 3, 2023. www.forbes.com.

Chapter Two: How AI Learns

14. Quoted in Sascha Brodsky, "Examining Synthetic Data: The Promise, Risks and Realities," IBM. www.ibm.com/think/insights/ai-synthetic-data.
15. Quoted in Mike Snyder, "Intel CEO Brian Krzanich's Embrace of the Future Snagged by Flaws of Past," *USA Today*, January 7, 2018. www.usatoday.com.
16. Quoted in Sarah Brown, "Machine Learning and Generative AI: What Are They Good for in 2025?," MIT Sloan School of Management, June 2, 2025. https://mitsloan.mit.edu.
17. Quoted in Richard S. Sutton and Andrew G. Barto, *Reinforcement Learning: An Introduction* (2nd ed). MIT Press, 2020, p. 1.
18. Quoted in Pablo Villalobos, "Will We Run Out of Data? Limits of LLM Scaling Based on Human-Generated Data," arXiv, June 4, 2024. https://arxiv.org.

Chapter Three: How AI Predicts Words

19. Quoted in Will Douglas Heaven, "Meet the New Biologists Treating LLMs Like Aliens," *MIT Technology Review*, January 12, 2026. www.technologyreview.com.
20. Quoted in Rick Merritt, "What Is a Transformer Model?," Nvidia, March 25, 2022. https://blogs.nvidia.com.
21. Quoted in Clara Davis, "Why Do AI Chatbots Keep Misunderstanding Sarcasm and Can We Fix It Without More Data?," Alibaba.com, February 3, 2026. www.alibaba.com.
22. Quoted in Monica Viggars, "Gender Bias in AI Systems—and How We Might Solve It," Medium, September 30, 2024. https://monicaviggars.medium.com.

Chapter Four: How AI Sees and Creates Images

23. Quoted in Kaia Glickman, "Computers Are Getting Much Better at Image Recognition," *Smithsonian*, November 3, 2025. www.smithsonianmag.com.
24. Quoted in Bella Isaacs-Thomas, "How AI Turns Text into Images," *PBS NewsHour*, January 12, 2023. www.pbs.org.
25. Quoted in Susan D'Agostino, "The Computing Pioneer Helping AI See," *Quanta Magazine*, October 24, 2023. www.quantamagazine.org.

26. Quoted in Andy Oliver, "Raging Against the Machine: Comics and Generative AI 'Art' Part 1—'Nothing It Offers Has Any Tangible Value or Originality,'" Broken Frontier, July 28, 2025. www.brokenfrontier.com.
27. Quoted in Oliver, "Raging Against the Machine."
28. Quoted in Isaacs-Thomas, "How AI Turns Text into Images."

Chapter Five: The Full Potential of AI

29. Quoted in Samuel Greengard, "Multimodal AI Connects the Digital Dots," *Communications of the ACM*, December 7, 2023. https://cacm.acm.org.
30. Quoted in Amit Chowdhry, "Sparkli: $5 Million Pre-Seed Funding Raised for a Multimodal AI-Native Learning Engine for Children," Pulse 2.0, January 24, 2026. https://pulse2.com.
31. Quoted in David Roe, "AI Adoption Is Increasing but Challenges Remain," CMSWire, January 31, 2019. www.cmswire.com.
32. Quoted in Susanna Ray, "What's Next in AI: 7 Trends to Watch in 2026," Microsoft, December 8, 2025. https://news.microsoft.com.
33. Quoted in Kathy Pretz, "Stop Calling Everything AI, Machine-Learning Pioneer Says," *IEEE Spectrum*, March 31, 2021. https://spectrum.ieee.org.
34. Quoted in Ray, "What's Next in AI."

FOR FURTHER RESEARCH

Books

John Allen, *Transformative Technology: Teens and AI*. ReferencePoint, 2026.

Emma Carlson Berne, *Artificial Intelligence: A Graphic History*. Lerner, 2026.

Jennifer Kaul, *The Potential of Artificial Intelligence*. ReferencePoint, 2025.

Stephanie Sammartino McPherson, *Future Forward: Harnessing Artificial Intelligence*. Lerner, 2026.

Internet Sources

Marcel Binz and Eric Schulz, "How Well Can a Computer Think?," *Science Journal for Kids*, November 2023. www.sciencejournalforkids.org.

Coursera Staff, "How Does AI Work? Basics to Know," Coursera, October 15, 2025. www.coursera.org.

Maurice Jakesch et al., "How Can We Tell Whether We Are Talking to a Computer or a Person?," *Science Journal for Kids*, March 2024. www.sciencejournalforkids.org.

Tom Norman, "Artificial Intelligence 101: A Beginner's Guide to the World of AI," Craft, May 17, 2023. www.craft.do.

Organizations and Websites

AI in Action, Technovation

www.technovation.org/ai-in-action

Technovation's AI in Action is a free, project-based curriculum that guides young people through identifying real community problems and building AI-powered apps to address them, with lessons on machine learning fundamentals and solution design.

HowAIWorks.ai

https://howaiworks.ai

HowAIWorks.ai is a free website that explains AI in simple, visual lessons. It uses examples, interactive tools, and a glossary of AI terms to show how AI works and what different AI models do. Free registration is required.

Machine Learning for Kids

https://machinelearningforkids.co.uk

Machine Learning for Kids is a free website that teaches students the basics of machine learning through hands-on projects. Students train simple AI models to recognize text, images, numbers, or sounds and use them in beginner-friendly coding tools like Scratch.

Teach Teens Computing, Raspberry Pi Foundation

https://training-hub.raspberrypi.org/en/courses/teach-teens-computing-machine-learning-and-ai

Teach Teens Computing by the Raspberry Pi Foundation offers this free online course that introduces teens to the basics of AI and machine learning. It explains how AI learns and lets students train simple models using hands-on activities and free tools. Free registration is required.

INDEX

PICTURE CREDITS

Cover: Sankai/iStock

5: Andrey_Popov/Shutterstock
9: PV productions/Shutterstock
10: A2942 Ingo Wagner/Deutsch Presse Agentur/Newscom
14: Jes2u.photo/Shutterstock
18: Pressmaster/Shutterstock
20: ORebrik/Shutterstock
23: Naeblys/iStock
27: Alexandre Rotenberg/Shutterstock
29: Kryazheva Alena/Shutterstock
33: Tatiana Gnuchykh/Shutterstock
37: evrymmnt/Shutterstock
38: Shutterstock AI
42: Prostock-studio/Shutterstock
46: Michael Vi/Shutterstock
50: pikselstock/Shutterstock
52: DC Studio/Shutterstock

ABOUT THE AUTHOR

Traci Pedersen has been writing for more than two decades, with thousands of published articles and a collection of science books for children. She lives in the Tampa Bay area of Florida and is the proud parent of three adult children.